Title: Monorails
R.L.: 1.8
PTS: 0.5
TST: 192557

AMAZING TRAINS
Monorails

by Christina Leighton

BELLWETHER MEDIA • MINNEAPOLIS, MN

Note to Librarians, Teachers, and Parents:

Blastoff! Readers are carefully developed by literacy experts and combine standards-based content with developmentally appropriate text.

Level 1 provides the most support through repetition of high-frequency words, light text, predictable sentence patterns, and strong visual support.

Level 2 offers early readers a bit more challenge through varied simple sentences, increased text load, and less repetition of high-frequency words.

Level 3 advances early-fluent readers toward fluency through increased text and concept load, less reliance on visuals, longer sentences, and more literary language.

Level 4 builds reading stamina by providing more text per page, increased use of punctuation, greater variation in sentence patterns, and increasingly challenging vocabulary.

Level 5 encourages children to move from "learning to read" to "reading to learn" by providing even more text, varied writing styles, and less familiar topics.

Whichever book is right for your reader, Blastoff! Readers are the perfect books to build confidence and encourage a love of reading that will last a lifetime!

This edition first published in 2018 by Bellwether Media, Inc.

No part of this publication may be reproduced in whole or in part without written permission of the publisher. For information regarding permission, write to Bellwether Media, Inc., Attention: Permissions Department, 5357 Penn Avenue South, Minneapolis, MN 55419.

Library of Congress Cataloging-in-Publication Data

Names: Leighton, Christina, author.
Title: Monorails / by Christina Leighton.
Description: Minneapolis, MN : Bellwether Media, Inc., [2018] | Series: Blastoff! Readers: Amazing Trains | Includes bibliographical references and index. | Audience: Age 5-8. | Audience: Grade K to 3.
Identifiers: LCCN 2016052935 (print) | LCCN 2017010229 (ebook) | ISBN 9781626176720 (hardcover : alk. paper) | ISBN 9781681034027 (ebook)
Subjects: LCSH: Monorail railroads–Juvenile literature.
Classification: LCC TF694 .L45 2018 (print) | LCC TF694 (ebook) | DDC 625.4/4-dc23
LC record available at https://lccn.loc.gov/2016052935

Text copyright © 2018 by Bellwether Media, Inc. BLASTOFF! READERS and associated logos are trademarks and/or registered trademarks of Bellwether Media, Inc. SCHOLASTIC, CHILDREN'S PRESS, and associated logos are trademarks and/or registered trademarks of Scholastic Inc.

Editor: Nathan Sommer   Designer: Lois Stanfield

Printed in the United States of America, North Mankato, MN.

Table of Contents

What Are Monorails?	4
Built for Height	8
Touring Trains	16
Glossary	22
To Learn More	23
Index	24

What Are Monorails?

Monorails are trains that travel on a single rail. They can move above land and water!

rail

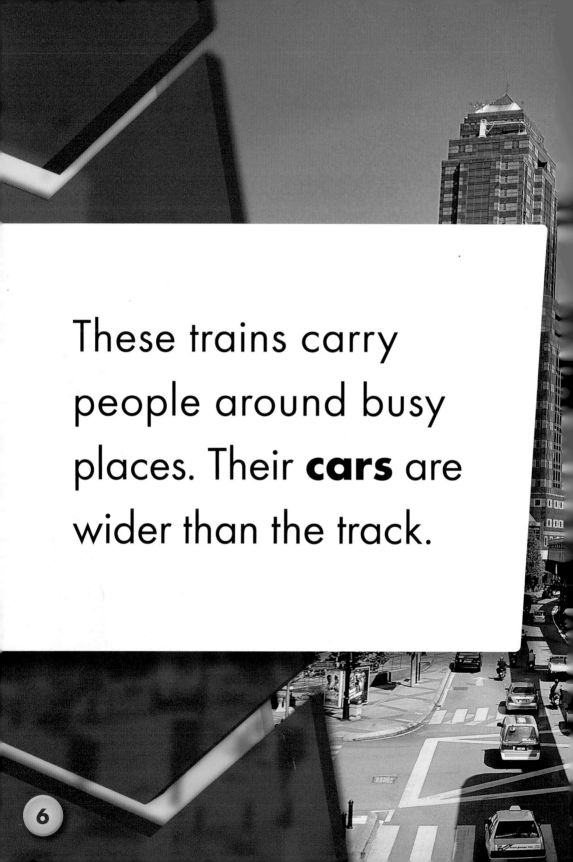

These trains carry people around busy places. Their **cars** are wider than the track.

Built for Height

Monorail tracks are often high up. **Passengers** enjoy looking at sights from the windows.

Monorails usually ride on top of rails. Some monorails hang from them!

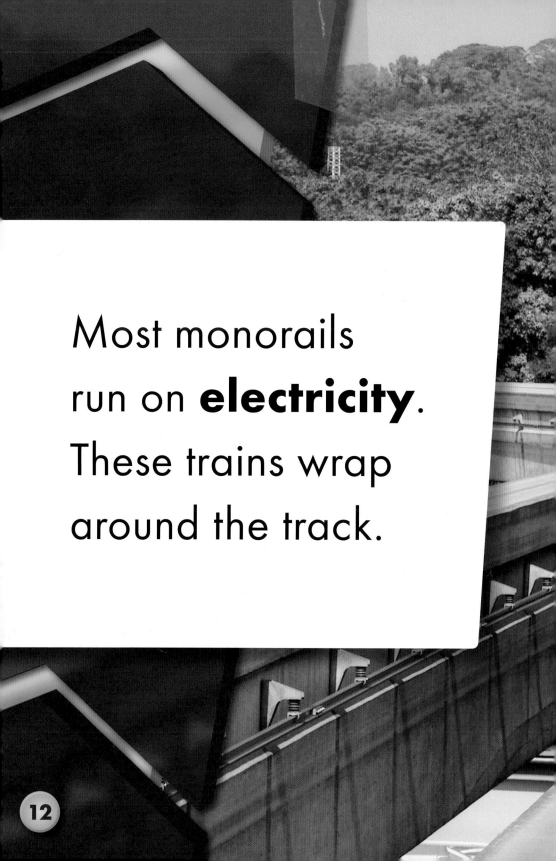

Most monorails run on **electricity**. These trains wrap around the track.

electric monorail

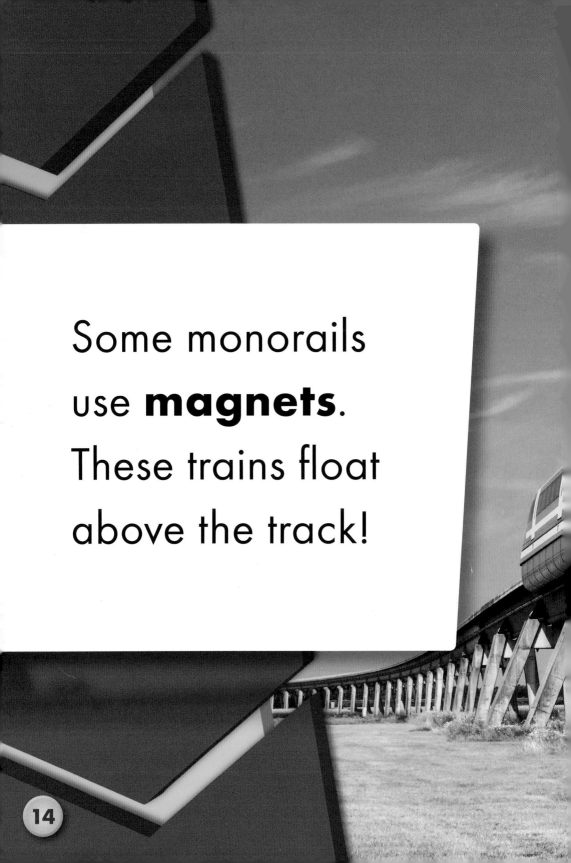

Some monorails use **magnets**. These trains float above the track!

magnetic monorail

Touring Trains

Monorails are usually found in big cities. They often help at airports.

airport monorail

Monorails also zip around **amusement parks**. Some even **tour** zoos!

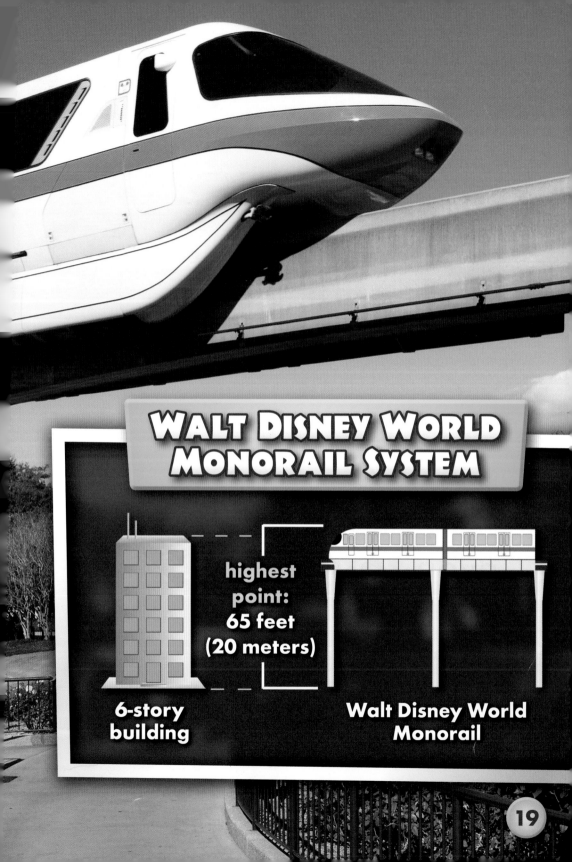

Walt Disney World Monorail System

6-story building

highest point: 65 feet (20 meters)

Walt Disney World Monorail

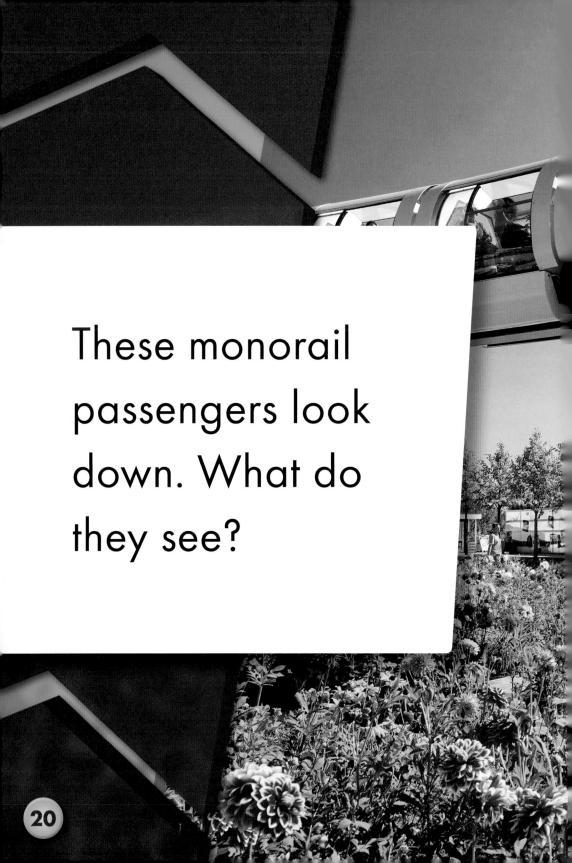

These monorail passengers look down. What do they see?

Glossary

amusement parks

places with games and rides

magnets

pieces of metal that attract other metals

cars

vehicles pulled by a train

passengers

people who ride a vehicle to get from one place to another

electricity

a form of energy that gives power

tour

to learn about and see a place

To Learn More

AT THE LIBRARY

Leighton, Christina. *High-Speed Trains.* Minneapolis, Minn.: Bellwether Media, 2018.

Phillip, Ryan. *Monorails.* New York, N.Y.: PowerKids Press, 2011.

Royston, Angela. *All About Magnetism.* Chicago, Ill.: Heinemann Raintree, 2016.

ON THE WEB

Learning more about monorails is as easy as 1, 2, 3.

1. Go to www.factsurfer.com.

2. Enter "monorails" into the search box.

3. Click the "Surf" button and you will see a list of related web sites.

With factsurfer.com, finding more information is just a click away.

Index

airports, 16, 17
amusement
  parks, 18, 19
cars, 6, 7
cities, 16
electricity, 12, 13
float, 14
hang, 10
land, 4
magnets, 14, 15
passengers, 8,
  20, 21
people, 6

rail, 4, 5, 10
tour, 18
tracks, 6, 7, 8,
  12, 14
water, 4
windows, 8
wrap, 12
zoos, 18

The images in this book are reproduced through the courtesy of: S. Forster/ Alamy, front cover; Scanrail1, pp. 2-3; imageBROKER/ Alamy, pp. 4-5, 20-21; Ravindran John Smith, pp. 6-7; Leonid Andronov, pp. 8-9; Hackenberg-Photo-Cologne/ Alamy, pp. 10-11; Elena Yakusheva, pp. 12-13; Bernd Mellmann/ Alamy, pp. 14-15; Oliver Hoffmann/ Alamy, pp. 16-17; Ian Dagnall Commercial Collection/ Alamy, pp. 18-19; number-one, p. 19 (building); nitinut380, p. 19 (monorail); Everything, p. 22 (top left); Pat_Hastings, p. 22 (top right); vanlhd_99, p. 22 (center left); Galkin Nikolai/ ZUMA Press/ Newscom, pp 22 (center right); Romas_Photo, p. 22 (bottom left); blickwinkel/ Alamy, p. 22 (bottom right).